Mohamed El Bahry

The pyramids are on the loose

All rights reserved. No part of this publication may be reproduced in any form or by any means without the permission of the author.

It is not allowed to use this publication in other media or in seminars, lectures etc. without permission of the author. The contents are carefully revised; however, without guarantee for correctness and completeness.

Copyright © 2019 Eng.: Mohamed El Bahry

Publishing and Production house:
Amazon-Kindle (Kindle-Create)

ISBN: 9781711969343

Pixabay License: All Pictures / Illustrations from Pixabay here in the Book are free for commercial use, no attribution required.

Thanks for Pixabay.

The biggest Dream of the Pyramids

One day, the pyramids wanted to travel around the world.

What? how could it be done?

The Pyramid Siblings

Pyramids are, just as all people in the world know, very old building. They have been built by ancient "Egyptians" Thousands of years ago and they still are now on their sites and receive daily visits from people from all over the world. This is the story of the pyramids siblings "Cheops (**Keo**), Keferen (**Kefe**) and Mekelenus (**Meke**)" and her friend the Sphinx (**Sphinky**).

Many have experienced the four together, good and bad, such as predators invade her and wanted to take possession of their treasures, as well as in present days, many admirers of these amazing buildings.

So, it was possible for them to hear almost all the languages of the world and to learn at the same time through their running visitors, who are also called tourists and their companions, known as the (tourist guides).

The Pyramids present themselves

- Hey, I'm **Keo** the biggest pyramid in the world, all right!
  And I'm **Kefe**, the second biggest pyramid.
- yes, hello and I'm **Meke**, the third biggest pyramid.
- Here I am, Sphinx, you can call me **Sphinky**.
- Hey, everybody, I am the Faraon Ramses the II, or you can call me **Ramsay**.

World Travel! how come?

One day, spoke the big brother Cheops (**Keo**) to his younger siblings, "**Hey,** may we want also visit these other countries and cultures, I am tired of always just here to stand around for millennia, **and** without a single bit of movement. **Finally,** I would like myself to travel around the world "What do you say?

At first Keferen (**Kefe**) was a little confused but finally he replied: **Well, I would also** like to see our family members, who are scattered somewhere in the world again, and you Mekelenus, what do you say?

Mekelenus (**Meke**) replied briefly: "I'm going to be with you,
I'm still young and want to learn a lot, and I can get some exercises".

Problems

At the end of conversation remarked Sphinky, the only one, that they had probably overlooked a problem and said: Hey guys you have a little problem!

Keo was astonished and asked, "What kind of problem Sphinky"?

> Sphinky replied, "well, have you ever thought about how big we are, and how should we to travel undetected by the little people through the world?

The Plan

Hemmm! Keo thought for a moment, and then spoke to the others about his plan, his plan, that he wished since a very long time.

The Professor

Keo said: I have heard a lot about a famous and very wise professor, he is to be a genius in many fields, and he can help us in our plan.

Sphinky was not very convinced by Keo's plan and asked him, "how we come by your professor to travel around the world".

Keo said: "I recently overheard a tour group that was talking about the professor that he really wants to visit us to learn more about the secrets of the pyramids. If he's here, I'll talk to him about our trip and the problem and I am very sure that he will help us.

Sphinky thought, "Oh great Keo, and if he is startled by your loud voice or even dies, what is it then! Perhaps, no one will come to visit us anymore? ".

Keo told her, that they all need not to have any worry, and he will take care quite well.

Moment of Awakening

One day a group of tourists came into the Cheops Pyramid, and while the guide the visitors led them into the King's Chamber to explain them that the pyramids were originally built as graves!

and he said again, unfortunately I forgot at the moment the name of the pharaoh of this Pyramid, but it is certain that the pyramids were tombs for the Pharaohs!

"Said a member of the group:" Now wait a minute!!! Please!
You do not know who this pyramid belongs to! But you claim that you are certain that they are tombs, in your opinion!
What do you want to tell us here of tombs and so on is totally nonsense "?

He tried to explain the fact that ...

At this moment, Cheops awoke from his sleep, and began to roar like a lion, AAAAAAhhhhhh bbbbbbbuuuuu.

The tourists were frightened and began to run out of the pyramid because they feared an earthquake, spirits or the collapse of the pyramid.

Only the man who had just tried to tell other visitors about the pyramids remained alone in their amazed and asked ": Where came this roar? "

Well, after all the people had left the pyramid and only one visitor was left, Keo knew that his hour had come, to which he had waited so long.

This is the man; he could only be the professor, who was expected for a long time.

His chance had come and he wanted to speak to the Professor, but oohh fright so excited denied him his voice, not the smallest sound could be heard. Quick Keo has closed the entrance of the king's chamber, so that the professor cannot even able to run away.

The professor had despite all the curiosity of great fear and said ":
what's happening now, I'm trapped, oohh God, how can I get out of here merely again, why did you do that with me, maybe I'm not such a good scientist, perhaps everything, what I have developed and invented, was not suitable for the humanity? "

Then finally found Keo his courage and his voice again, and he tried to speak relative quiet. Do not worry, Professor.

The professor believed at that moment to be crazy, a voice, a voice indeed, is the spirit of the builder of this pyramid, or is it God, who speaks to me?

God is that you? why have you trapped me here, please let me out of here, please?

At this moment Keo could resist a little laugh hardly, wait a minute he spoke (Keo): "Professor, I'm not God, I am Cheops, the pyramid itself, and I have very long time waiting for you to talk to you.

What! The Pyramid speaks! That's impossible, I am imagining now however a mere, the professor was talking more to himself.

Yes, it's me, I am the Cheops pyramid and you may call me Keo. I want to welcome you and talk to you, however please take a moment and take a place please.

The professor still could not believe he was really talking to a pyramid, this is all not just a dream? Nevertheless, he replied, with a little scared, "place! where is here however is except of the ground to sit on, no other place? "

Cheops now began to speak calmly and clearly: "Now sir, why I want to talk to you, is not so easy to explain. My brothers and sisters, the Sphinx and I would like to take a journey around the world.

What? the professor said, yes, yes, I still dreaming, ohhh I do not know how this dream ends, well, maybe I'm already dead, or was dying, ohhh I think I have a fever, but the dead do not have a fever, ohh yeh what is happened to me? and where are the other members of the group? Maybe they are also dead, and everyone speaks with the pyramid like me oohh oohh yeh yeh ...
and further pyramid Keo, what do you want to tell me!

Keo, said: We have other relatives, as you probably know, who are scattered all over the world, we finally would like to visit them, and learn about cultures of other countries. How all the people actually living in their countries who come to visit us every day for years and years? How it looks with them? We just know the desert.

I also think of my little brother Mekelenos, he should learn a lot and move around.
I also thought that you could join us, your knowledge and experience you have acquired all over the world would be so helpful to us and we would not be alone on our adventurous journey.

Well, Professor, what do you think of this idea?

The professor did not know, what just he has to say.
But then he said, "that's quite a crazy thing, millennia old buildings who want to travel around the world?
Tse, tse, tse, and how do you imagine the whole? We take the possibility of moving only as an example.
With your size, there are absolutely no suitable means of transport that can take you around the world.
What about the empty places that you leave here in Egypt? And what about all the people in different countries, you will meet?

At your current size, it is very scary for people. And, and, and ....

The Secret of the Magic Chamber

Keo answered the Professor quite calm and composed, that is here in the pyramid a mystery and magic chamber.

In this room, many mysterious materials are hidden with which or from which this journey can certainly realize with the help of the professor. In due time, we will open and show you this room.

He He ... never mind, the professor spoke to himself, magic room! and said, but the Egyptologists have not found something like that, oh well, any way let us have a look, how it goes on, Ok.

This answer made the professor naturally curious, because after all he even wanted to explore the secrets of the pyramids.

After a short pause, he said Keo, that he agrees with a common journey and offered him his help.

Keo was thrilled at the many adventures on which they would all wait.

##

After a little break called Keo: "Hey, Professor, do you know what I am most happy?"

Professor: No Keo. So, what, then?

On the Flying.
What are you talking about Keo?
Well when we all then high as the birds flying up in the sky through the air.

What? I think I did not understand you correctly! Did you say Flying!
Yes, Prof. of course, is exactly what I have said, we want to fly.

That surely cannot be a reality, I must still be dreaming, are there anyone cannot even wake me and get me out here.

What please, do you want with your million tons of weight and as such a huge building so easy to fly through the air? Do you think you can fool me that easy?

Not at all Prof. we have our flight obelisks, as well as in your country, the witches flying with their broomsticks, so we fly, pyramids, from the Orient up with obelisks, which are much more beautiful and more stable than the broom, and most importantly they are in its front as sharp as the rockets today.

Professor said: Oh God, that may indeed be funny, with obelisks, we will fly high up in the air, I believe all these things simply not, after all I am a man of science and not a magician!

Then, Keo has unlocked the gate and told the professor: well Professor, you can go home now, I've opened the gate, and think about our journey in the calmness, and please do not tell anyone about it.

The Professor (Kotomoto) walked back slowly to his hotel which lay in Cairo, He felt incredibly tired and exhausted, so first he had to lie down. In his bed he thought again about everything what he had just experienced until recently.
He spoke with an incredibly huge pyramid that would like to travel around the world with him and the other pyramids, and even the sphinx on the obelisk.

A dream or a fever or perhaps the curse of the pharaohs must have packed him in this pyramid and however everything was so really been for him.

After he had slept a little, he then sat down at his table, which stood against the wall and opposite of his bed, and began to create a travel plan for himself and his companions.
On what he would have to think about all things,

and of course also about how much victuals he should plan for himself.

Finally, he was the only human being in this incredibly crazy tour group.

Unexpected Visitor

At night, happened at the Pyramids something very surprising. An unexpected visitor joined them.

Hello, hello you, it called from below up to them! How's my pyramids and of course my beloved Sphinx? Are you sitting still here at this place comfortable?

The one who asked this question was the great Ramses himself, and coughed while talking to them so terrible.
Quickly he took something that could not be seen to the other, out of his pocket, and put it into his mouth. Well, the great Ramses actually suffered under an asthma attack.

The pyramid siblings were very surprised that Ramses had left his rightful place in downtown Cairo to come visiting them in the desert.
At first, they were speechless and got out no sound.

Hi Ramses, Keo said after quite a while, what have you lost here with us in the desert?

Are you sick or what has caused you to leave your place in Cairo? Why are you after so many years back to us?

Yes Cheops, I became ill, by the so-called industries of the people, there are no more shipments that are performed with horses.

Everywhere, driving there, only those noisy and smelly metal boxes through the area, and what they give from themselves, people call it exhaust, which smell not good at all, and over the last years I got thereof asthma.

I stood in the middle of Cairo and could not move from the square.

I wanted a long time to escape from there but these weird metal boxes have always stopped and prevented me from escaping from there.

One evening, however, probably a few smart people in Cairo had some understanding, and the government induced me mop of this terrible place, back to the desert, where there is still reasonably clean air.

They have finally realized that this horrible dirty air attacks my skin, and feared that I might no longer could be there in a few years, so I am now be turned after so many years again tonight back to you.

Oh Cheops, I'm so glad to have gotten away from this chaotic place in the middle of Cairo.

I am incredibly glad to be with you again.

We are also extremely pleased that you have come here to us ... but oh eeehh.

Yes Keo, what is it? Speak please.
Oh Ramses, I would like to inform you something, maybe you do not like it,

What then Cheops? Don't you want me here with you?

No, no Ramses it's not so, but we have planned something.
Meke, Kefe, Sphinky and I want to make a great world travel, and about you, we have, of course, did not thought about your engagement.

we did not know about you Ramses, so I mean, we have not considered you while making our plan.

What? a trip around the world? Oh God, no, I do not want to travel again, I do not want need somewhere to go, and there again breathe this nasty exhaust: said Ramses.

In any case, I would not be there, I want to stay alone for a long while. I'm sick enough of the crowds and spectators, here in the desert it should to me going back better than I had all the long years before.

Oh Ramses, I am really sorry, but even here there are loads of tourists and spectators who

21

visit us from morning to night, touch and goggle that you can hardly imagine. Said Meke.

In any case, you can find a little more rest and some clean air than used as in Cairo, at least at night. Said Sphinky.

You can follow your plan quietly, I will not stop you.

I came here to stay, and I do not intend to change everything again.

Keo has briefly consulted with the others about the new situation, and all of them could fully understand that Ramses had no desire to go there again somewhere else.

When it was day, Professor Kotomoto came back to the pyramids.

He told them about it, that he had concocted a new plan, what their travel times concerned.

He suggested that they travel only in the late evening and at night, so that they could all be back in their places the next morning without their visitors would ever notice, they had left their places.

Professor Kotomoto told the pyramid siblings "Every time we fly, we fly to a different place, then return to morning and make a break here during the day to rest and relax and the next or the other evening, we will fly off again ...

Keo interrupted the Professor and said hey Professor we have got here last night a very important state visit we would like to introduce to you.

It is Pharaoh Ramses he became very ill in Cairo and could no longer endure there. He decided from now on forever here to stay with us.

Oh God, no, not one more, then I must recalculate everything again ... said Professor Kotomoto.
No, no Professor do not worry, he will stay here at the place, he does not want to travel with us. Said Sphinky.

Well is also better that way. Said Professor Kotomoto.

I will keep here the "status" Ramses said, so I mean, I'm going to monitor this space here, and should unexpected something happen, then I can however be in contact with you.

Ha ha ha ...What? How then Ramses? How was that all go, we are not even just so disappeared around the corner: Asked him the Professor.

Now Prof. as you know nowadays there is so little funny things or devices, they are also called mobile phones, in our times there was nothing like that.

I've got me one of these beautiful devices. You can of course also have such devices, so we can keep in touch, what do you think about my idea people? I've always had good ideas, hey (smiles Ramses), is not Sphinky?

Yes, super Ramses, which is a good idea, I would like to have a cell phone, said the small Meke.

Professor Kotomoto: Well convinced Ramses. So, I will go after, and we will get such talkies, even though I do not like this kind of frightening devices in my pocket.

But Ramses was right you can never know if we do not maybe get into a difficult brisk threatening or dangerous situation. But I have one more question, how fast we can then at all fly with your obelisk?

Sphinky answered him fast enough, Professor, do not worry, our obelisks are very fast, I think, when the broom fly, then our obelisks can also fly.

What: This statement has surprised the Professor, you do not know if the obelisk flying and compare it with the broom !! and plan a world journey? Oh Got ... said the Professor.

Prof. Kotomoto asked Sphinky, if they would be then ever tested and flew it somewhere?

Sphinky answered: No, but I've watched on the television such flight boxes.

Prof.: What, on television, where did you have a TV, I see only desert sand and stones?

Well, sometimes when everyone is asleep, I sneak here like a cat so slowly around, sometimes I go down into the village and find me there a quiet place from where no one can see me. There are Cafes where people sit and play cards or chess there, and some smoke and drink even just coffee or tea and a talk. Since then I sit around and watch everyone so quiet, yes and sometimes I stare then also a little bit television, quite incidentally.

Prof. Kotomoto answered, you know, every time you surprising me more and more. I will simply trust when we all work together, then we will do it too.

Return, I cannot anyway and besides, it is also for me as a scientist, a unique opportunity even though I do not know exactly if I all just dreaming or not. any case I hope not this is all just a dream. I'm going to do now on the way to meet all other arrangements for our first Journey. We will meet again in the evening, I'm very curious about the magic chamber.

After the Prof. had done the remaining things, he returned to his hotel room and began soundly to sleep. In the meantime, broke out between Keo and Sphinky a discussion of the flying machines. "Which machine do you have seen on television, he asked Sphinky?"

Well, she replied, such great balloons that rise with hot air upwards and also so funny oblong shaped airships.

Keo answered but Sphinky ". but Sphinky, with that we cannot fly as fast as we need to fly in one night, after all, we still have only the nights available".

In the evening came back Prof. Kotomoto and showed his tour group brand new mobile phones, which he had bought to her great surprise. He asked Keo whether he was finally able to enter into the magic chamber, after he finally wanted to know what materials are hidden in it and what he needed for their journey and could use.

Keo unfortunately had to disappoint him ... No, no Prof. unfortunately is not so simple, it opens only when we need something otherwise it remains closed. Assuming without an important reason into the chamber, it could be that they are no longer open, so be careful Prof.

All right Keo but I would like to inquire about what provide for opportunities for us, we are going to make unusual journeys.

Well, Keo said, but once you have to tell me what you need, then I'll tell you the password and you can enter.

By the way, Keo, I have another little big problem, I find it incredibly difficult to discuss and entertain with all of you, because of your size contrary to me, the relationship between your range of motion and my is however very different. Do you have a solution for that Keo?

Keo: Sure, of course I have, do you need them now or later, Professor?
Prof.: Oh God, why you do not answer such directly, in any case, that's wonderful, do you really have a suitable solution for this?

Keo: As I said before, do you need the solution now? It is limited in time, should we use it without really need to at certain moment, we could thus provoke a real problem.

As I have told you before in the beginning, there is always time for a problem a suitable solution, this knowledge and insight we owe to the ancient Egyptians. They were in their development and their inventions much further ahead than people believe.

I will dedicate you in due time to all those secrets, and I really hope that you will also be able to conceive and understand everything.

There are a lot of secret knowledge, and the people are not ready to understand everything.

I must say you are very impatient Professor also very curious, I do not mean that negatively, no, quite the contrary, it is positive for you, and it proves to me that you are the brilliant scientists, for whom I consider you. The ancient Egyptians invented many things, such inventions learn with me but as I said in time.

Thank you Keo answered the Prof.: I am honored much, that you have such a high opinion of me.

Language Pills

Hey Keo ... there is something that I must discus with you, in our travels we will certainly coming into contact with the people of the country, in which we travel, Take, for example, China, I cannot speak Chinese, what could we do then?

Is that a problem for you Prof., asked him Keo? Prof.: Well, we'll have to expect that people will want to talk to us, we are not exactly what could be described as normal tourists.

OK professor Kotomoto, I'll tell you something, there is in the magic chamber so called language pills, I hope you are now a little calmer in this regard.

Prof.: What Language pills? What does that mean?

Keo: Well, the ancient Egyptians developed these pills from certain materials, you just need the right pill of the the respective language dissolve in a glass of water and drink and in a few minutes, you will understand every word of that language.

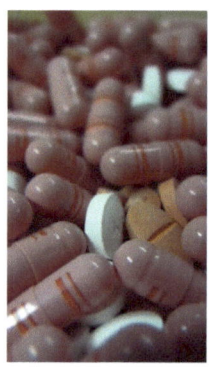

Prof. Keo ... that's incredibly fascinating, please tell me more how the Egyptians did that. if you want to speak a particular language, taking pill for that language, the ancient Egyptians had developed such pills from certain resources mixed with certain area of the brain of the dead, for example, if a foreign die bought the Egyptians his brain to them to conduct experiments, and thus they have developed such pills, but...

Prof.: But ... what Keo? You're making it harder for me ...
Keo: Well there's always the side effects.
Prof.: even also!!! ... what kind of side effects, you're scaring me thereof Keo?

Keo: Well, suppose that the pill came from a brain of a thief, then it could be that his behavior influences on the property of the pill, and so on.

Oohh God, what adventures are waiting for me.

I will give you only a short overview, because it is not completely safe to use these pills.

Everything I do not quite know they have different materials mixed together, because they were familiar very well with the mummification of the corpses, they used certain areas in the brain that are responsible for language, inserted them into the language pills.

Then, for example, if a foreigner was mummified, so you could insert his mother tongue so it happens that you can understand all languages and speak it also.

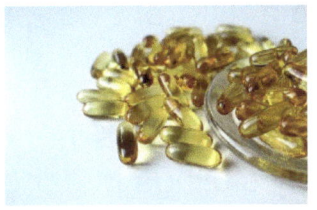

Unfortunately, they are not succeeded certain properties -to exclude- of persons, who were preserved for eternity, so, it can be as a side effect, quite possible that these properties are transferred to the person, who get the respective language pill, so might prefer, for example, get a pill that comes from a former thief, or liar, etc.., such it is not excluded that you Mr. Professor could possibly steal something.

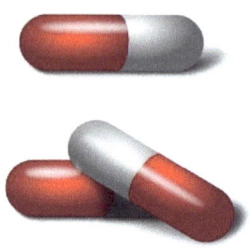

So, it is quite possible that the properties of the pill can affect your behavior.

also, you know that the effect is timely limited so as vitamin pill or something like that.

Oh my God Keo on what I have just admitted myself, what adventures we will go towards, what will happen on to us, you must promise to support me in any case, help me with your knowledge, otherwise I may be change my mind perhaps in a different way, can and will you promise me that Keo?

Keo answered him, "Sir, it is not my intention to bring you and my siblings in danger and of course I will do everything to ensure that any of our group is harmed, I promise you here and now.

So, they all spent together quietly for a while before they re-arranged to meet the next day.

The Miracle of the Numbers

Prof. Kotomoto asked Keo, how the old Egyptians have worked with the numbers and mathematics?
Then Keo asked him, Prof. do you know the difference between the numbers and why they are look like that?
What do you mean Keo? Prof. responded.
Why is the form of the number FIVE looks like this 5 and not like 9 for example !??, Keo replied.
Aaahhhh ... was the reaction of Prof. Kotomoto!! Ne, I really do not know! Kotomoto said.

Keo told him, well then ... now you have a reason to admit the chamber, you can go into the magic chamber and borrow the book "The Wonder of numbers / or / The Logic of the Numbers", read the book, then you will find the answer for this question, go in there and call "The Miracle of Numbers" as the password, then you get the book.
The Professor was very happy because he can go into the magic chamber at last, he walked into the magical chamber and was a little anxious and called out the password, then

opened a door and saw an extensive library and was very surprised with opened mouth, then came to him a pickup tray and flew opened and was in it the book, the Professor took the book and walked out of the magic chamber such as from a dream.

Then Keo asked the Professor, hey Professor Do you have the book, and seen the magic chamber?
Yes... Yes ... Keo ... it is incredibly unbelievable ... I cannot describe it ... what was in it? I cannot believe what I saw there.
Ha Ha Ha ... answered Keo, now you know Professor ... I already told you ... this is not normal for you, you go and sleep and tomorrow read the book.
Professor Kotomoto went to the Hotel and tryed to sleep, but he hast he book, he desided to start reading it.

Part 2, The Logic in the Numbers

Mohamed El Bahry

The Logic, Evolution and Appearance of the Numbers

AUTHOR

Mohamed El Bahry,
B. Sc. Degree in Shipbuilding
Engineering, 1980
Born at 24.11.1955 in Egypt
Hometowns: Berlin, Germany, and
Port Said at the Suez Canal, Egypt,
Northeast point of Africa

The Logic, Evolution and Appearance of the Numbers

- **The Principle of the angles and the Logic of the Numbers**

- **The Logic of Number TEN (10) in comparison with (X) and the new Number (8)**

Bibliography information of the German Library:

The German library registers this publication in the German National bibliography; detailed bibliographical data are available on the Internet at http://dnb.ddb.de

All rights reserved. No part of this publication may be reproduced in any form or by any means without the permission of the author.

It is not allowed to use this publication in other media or in seminars, lectures etc. without permission of the author. The contents are carefully revised; however, without guarantee for correctness and completeness.

Copyright © Mohamed El Bahry, first published on 2008 & Now 2019

- Art Works: Christel Koch-El Bahry
- Idea, Conception, Design and Illustrations: Mohamed El Bahry
- Publishing and Production house: (Now) Amazon-Kindle

TABLE OF CONTENTS

The Biggest Dream of the Pyramids 3
One day, the pyramids wanted to travel around the world .. 3
What? how could it be done? 3
The Pyramid Siblings 3
The Pyramids present themselves 4
World Travel! how come? 5
Problems 6
The Plan 6
The Professor 7
Moment of Awakening 8
The Secret of the Magic Chamber 14
Flying! 15
Unexpected Visitor 18
Language Pills 29
The Miracle of the Numbers 33
Part 2, The Logic in the Numbers 35
History of the Book 42
Introduktion 44
Chapter One (I) The Roman Numbers: I, V, X, L, C, D and M 47
Table of Roman and Arabic Numbers 48
The Logic of the Roman Numbers 49

The Roman number One 51

The Roman number Two 53

The Roman number Three 53

The Roman number Four 54

The Roman number Five 56

The Roman number Six 56

The Roman number Seven 58

The Roman number Eight 60

The Roman number Nine 60
The Roman number Ten 61
The Roman number Eleven 62
The Roman number Twelve 63
The Roman number Fifty 64
The Roman number One Hundred 65
The Roman number Five Hundred 66
The Roman number One Thousand 66

CHAPTER TWO (2) THE ARABIC NUMBERS 69

The Angle Principle and the Logic of the Arabic Numbers 70
The Angle 71
Training example 72
The Arabic number One 72
The Arabic number Two 74
The Arabic number Three 75
The Arabic number Four 75
The Arabic number Five 76
The Arabic number Six 78
The Arabic number Seven 79

The Arabic number Eight "2nd Form" 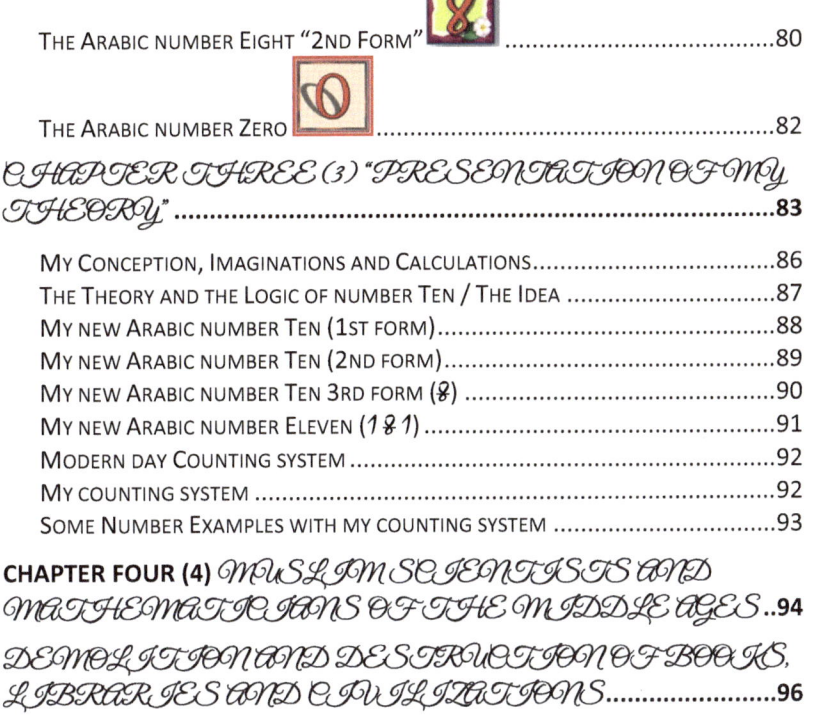 80

The Arabic number Zero 82

CHAPTER THREE (3) "PRESENTATION OF MY THEORY" 83

My Conception, Imaginations and Calculations 86
The Theory and the Logic of number Ten / The Idea 87
My new Arabic number Ten (1st form) 88
My new Arabic number Ten (2nd form) 89
My new Arabic number Ten 3rd form (8) 90
My new Arabic number Eleven (1 8 1) 91
Modern day Counting system 92
My counting system 92
Some Number Examples with my counting system 93

CHAPTER FOUR (4) MUSLIM SCIENTISTS AND MATHEMATICIANS OF THE MIDDLE AGES .. **94**

DEMOLITION AND DESTRUCTION OF BOOKS, LIBRARIES AND CIVILIZATIONS 96

THANKS 97

CONCLUSION 99

41

History of the Book

The reason which convinced me to write this book has a very funny background.

For a long time, I did not know that these numbers (0123456789), which determine nearly all of our daily routines, were Arabic numbers. In Egypt, my homeland, these numbers were always portrayed as "European numbers".

My first journey to Europe in the year 1977 led me to Paris (France). I went there into a mosque to pray, and I read pages from the Koran and astonishingly noticed, that the text and pages were numbered with these "European numbers".

I was extremely surprised and curious to find out why the Koran was provided in another country with such foreign numbers.

Many years later, approximately 1991, I visited the University of Trier in Germany to learn the German language. Then One day I picked up a mailing ticket from the post office, where upon I read a sentence, „the provided data and information are allowed only with Arabic numbers". I was then a little bit confused and which numbers were meant with that?

Then, it became clear to me that these numbers, which we use every day and everywhere, are the original Arabic numbers.

Therefore, I decided to take it upon myself to research and find the answers to my unanswered questions. How were they actually developed? Who developed, modified and designed them? Why do we still use them today?

In 1997, I began to study and became a great friend of the computer science. Thus, it came into my mind to start writing about and illustrating the numbers as we know them today and their logics. Now it is possible for me to explain in detail and show in beautiful diagrams the history, the logic and the Origin of the numbers.

Without my efforts and research in trying to find out about the origin, appearance and the logic of the numbers written in this book, I would never come across the idea, to modify the number Ten (10), and to present a new number system. In addition, and maybe One day I could say;

"I may have extended and reformed a very old numerical system"

I hope that this book with its contents, techniques and descriptions will open your minds to the fascinating world of numbers, and thank everybody for taking the time to read my book. **"I hope you have an Interesting journey!"**.

Introduktion

Many Thanks to my Wife Christel Koch-El Bahry; I used Art Works done by her, and many thanks to my friends for corrections of the text.

Thanks to the teachers of my daughters, that gave me the courage to publish this book. Once I asked them whether they would know as teachers how the numbers were developed! And who modified them?

The answers surprised me; neither they nor any other another person, who I asked, gave the right answer about the Origin, Appearance, Logic and the Development of the numbers.

Supposing you shall and want to explain and make the mathematical homework understandable to your children, that should not be a problem as you have taken your personal knowledge and combined it with the experience from your teachers, and then exactly at this moment you have forgotten One critical question.

Do we really know whether everything is exactly like our teachers have taught and explained to us?

The ability of the Intelligence is a process, which grows through and with many practical experiences.

But what can we do to give our children a better understanding of mathematics so that they can understand the logical meaning and the composition or structure of mathematical operations?

As a father or mother, you would be familiar with the situation when mathematics is studied at home, and the children's tears flow because of such silly mathematical equations.

All of us, children and parents alike are upset that the child must learn something now that we have already learned when we ourselves were that age, and they can not understand it because there logical thinking does not accept it. Then the following questions can arise, where do these silly numbers and this counting come from?

Perhaps you can find a good excuse or even know a right answer.

I came to a surprising conclusion from the attempt to find an answer to this question:

Where do our numbers and mathematics come from?

This absurd question was the question that I asked many Teachers, Scientists and people from different society levels. The answers that I received were not satisfactory to me. While I already knew that the number (0) ZERO originally came from the Arabian region.

This was very disappointing for me to realize that with mathematical formulae many complicated tasks are able to be solved, but to find an answer to such a simple question like (how and where numbers come from), was impossible. Concerning Information About the history of mathematics One can find a lot of useable data on the Internet, which are repeated everywhere, but finding an answer the core question "Where do our numbers come from, her Design, Origin and Evolution "there was not possible.

Someday I come across a German sentence the following; "Bücher, die einem zu fehlen scheinen, muss man selber schreiben" (Jean-Paul).

It means: "when you think a book is missing, or if you missed a Book, then write it yourself".

According to historical and verbal information, it is most likely that the hand and angle played a very essential role in counting operations which is a rough answer to my core question on mathematics.

So, herewith I would like to express my assumptions, thoughts and theories about the logic, appearance and history of the Arabic and Roman numbers, as well as their supposed origin with this small and quite descriptive book that is suitable for children and pupils alike. I hope that this book meets the requirements and expectations of every reader and wish you all fun when you investigate on my book the Empire of

"The Logic, Evolution and Appearance of the Numbers".

Chapter One (I)
The Roman Numbers: I, V, X, L, C, D and M

The Roman numbers (I, V, X, L, C, D and M) are still used today more than 3 millennia after their creation. A disadvantage of the Roman system is that it is not suitable for fast written calculations.

As long as we do not know the right history of the Roman numbers, everyone may have the possibility to express his assumptions and theories for the origination of the numbers.

In this chapter I would like to show my assumptions on how the Roman numbers could have been developed.

My thought is that the Hands and Fingers show the logic of the Roman numbers and their origin. Instead of using the whole hand or their fingers to paint or write, the fingers were used for counting, using only the Two final fingers (thumbs and little fingers), thus the result was the symbols I, V and X.

After, they possibly developed other roman letters to make it easier to represent larger forms and quantities, and finding that these new letters were harder to display using only the fingers of Two hands. Thus, these forms or letter „L, C, D and M" were modified by using arms and body.

Table of Roman and Arabic Numbers

From Zero to Twelve	From Ten to One Thousand
Number Designation = Roman Number = Arabic Number	
Zero = it doesn't exist = 0	Ten = X = 10
One = I = 1	Twenty = XX = 20
Two = II = 2	Thirty = XXX = 30
Three = III = 3	Forty = XL = 40
Four = IV = 4	Fifty = L = 50
Five = V = 5	One Hundred = C = 100
Six = VI = 6	Five Hundred = D = 500
Seven = VII = 7	One Thousand = M = 1000
Eight = VIII = 8	The number Zero doesn't exist in the Roman number system.
Nine = IX = 9	
Ten = X = 10	
Eleven = XI = 11	
Twelve = XII = 12	

The Logic of the Roman Numbers

Before and since many thousands of years the communication among people existed with the hands. The hand as always been, and will most probably always be the most important function that we as human beings possess, especially for our children who use the hands and fingers when they begin to learn and count. The quantities for example of animals, food and groceries were always indicated, counted or computed between humans by hand movements. Therefore, it is recognized that ONE finger represents One Unit, the whole hand as five units and both hands as TEN Units, or one could say; the hands are used as an official arithmetic and logic unit.
(ONE = I), (FIVE = V) and (TEN = X) represented the beginning and the principle of the computations at that time.
One imagined number Four in such a way that number Five would be displayed with the hand, thus the whole hand would be shown and by hiding One finger, thus surrounded behind the palm there were only Four fingers optically visible. Number Four was born after number Five,
It means: the hand (-) the thumb = Four fingers.

Calculating number Four in the Roman system:
V-I=> the thumbs (I) taken off behind of the hand (V) = IV

Calculating number Six: V + I = VI

The left or right hand + the thumbs of the other appropriate hand = FIVE + ONE finger equal SIX fingers.

The number Nine is computed and originates in the same principle as the number FOUR: X-I=IX. One shows both hands and then puts One finger (the thumb) behind the palm and then number Nine optically visible.

Calculating of number Eleven: X + I = XI

It is more difficult with number Eleven, because it depends here on the concentration of the opposite person.

At first one shows both hands and directly afterwards One finger of One hand, therefore One shows first number Ten and additionally picks up again One finger of One hand, and so number Eleven is visible. However, if the person on the opposite side should not have watched out, they could also think, that only number One was displayed because they missed the two raised hands before. Therefore, this system begins to become more difficult and confusing.

Hand counting (I, II, III, IV, V, X)

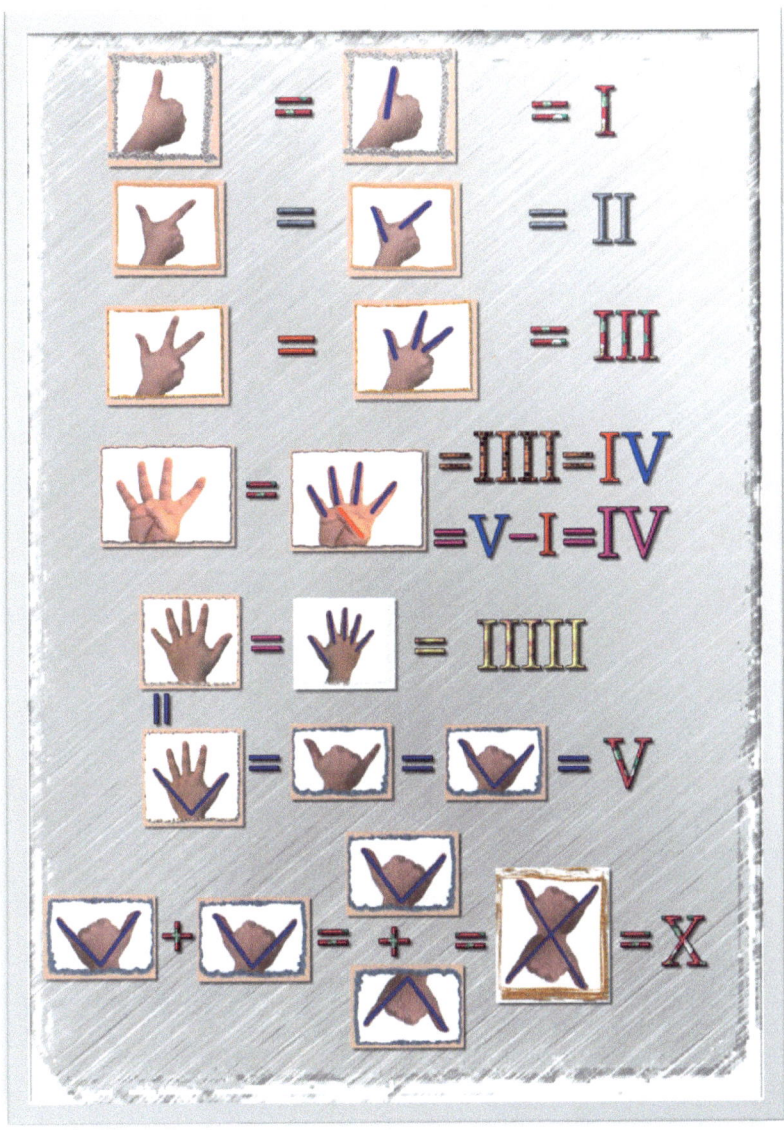

The Roman number One

The Romans used the thumb to display the visible sign which means a quantity of ONE unit, for example ONE picture.

One Picture = I Unit

The Roman number Two

The Romans had begun their counting with the thumb then with the following fingers of the hand, thus index finger, middle finger etc... Starting from that, the result then was number Two, the thumb plus index finger.

Two Pictures = II Units

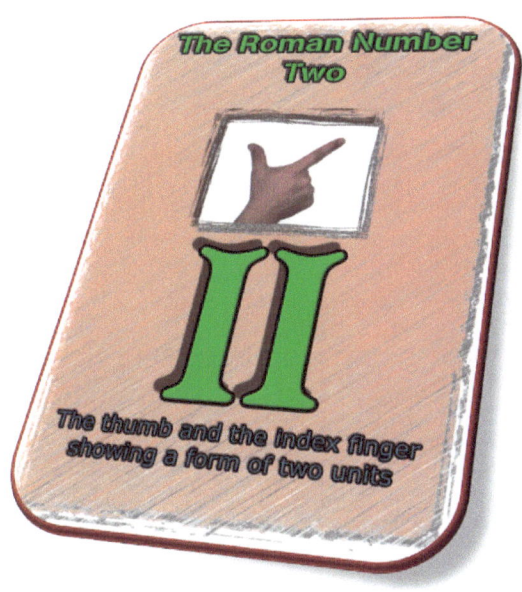

The Roman number Three

Like the next graphic, the Romans wrote their number Three and they represented it with THREE fingers of one hand.

Three Pictures = III Units

The Roman number Four

The Romans have followed their logical way of counting in the case of number Four. They first began with Five fingers (showing the whole hand) and then hiding One finger behind the palm to reach number Four. Now number Four is visible as a mathematical formula, **five** minus **one** is equal to **four** in the Roman way of counting.

Four Pictures = IV Units

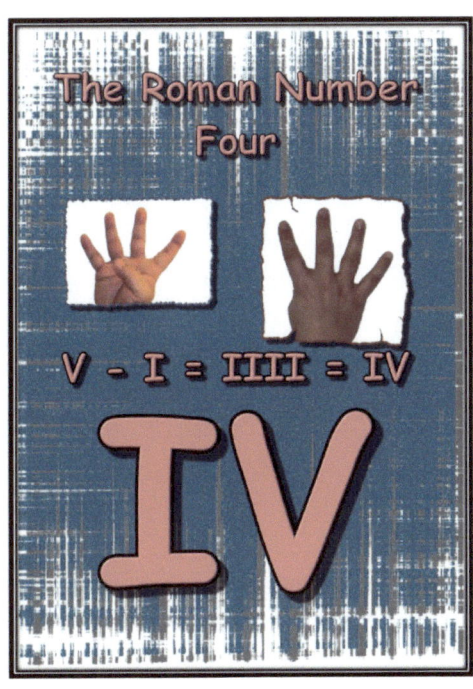

The Roman number Five

The assumption of number **Five** is based on representing the visibility, logic and principle of the Roman numbers. They most probably represented number **Five** by showing the whole hand and so used as a symbol **V**.

In general, the thumb and the little finger show the Form (**V**) by **two** connected lines.

Five Pictures = V Units

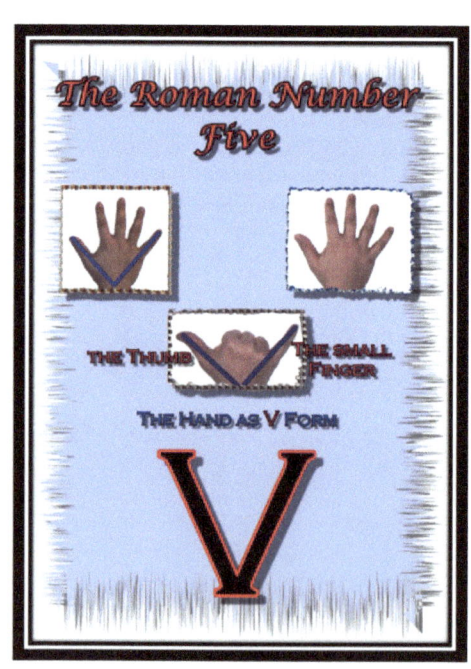

The Roman number Six

In order to remain with the principles of the Romans, they began directly with **FIVE** fingers of **One hand**, thus showing a visible number Five and at the same time by showing **the thumb** with the other hand number **One** which gives the mathematical equation:
FIVE + ONE = SIX
Therefore, displaying a visible number **Six**

Six Pictures = VI Units

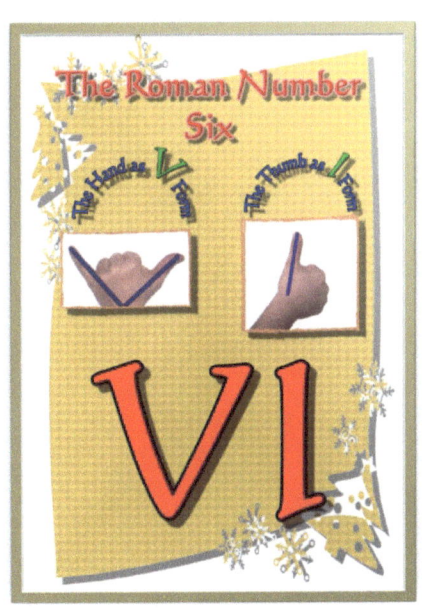

The Roman number Seven

For the Roman number Seven, one uses just as for the number six, first the entire fingers of one hand, it means Five fingers, and then Two further fingers of the second free hand were added.

Seven Pictures = VII Units

The Roman number Eight

Here are all five fingers of the left hand and additionally three fingers of the right hand can be seen, which represents the form for Eight Units or number Eight.

Eight Pictures = VIII Units

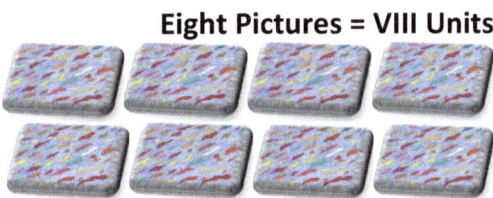

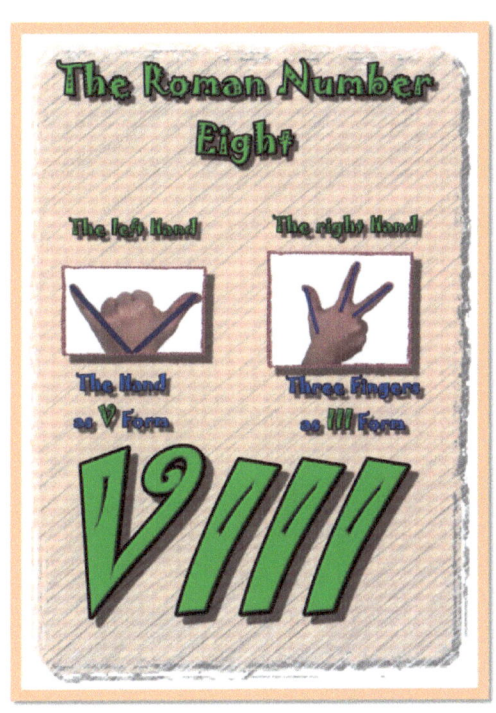

The Roman number Nine

Number Nine was surely developed in the same logical and mathematical computation of number Four.
By showing both hands with all ten fingers, and then hiding one finger, e.g. setting one finger behind the palm to achieve the result which is now clearly a visible number nine. Thus, numbers four and nine are regarded as mathematical formulas:
FIVE minus **ONE** equals **FOUR**, and **TEN** minus **ONE** equal **NINE**.

Nine Pictures = IX Units

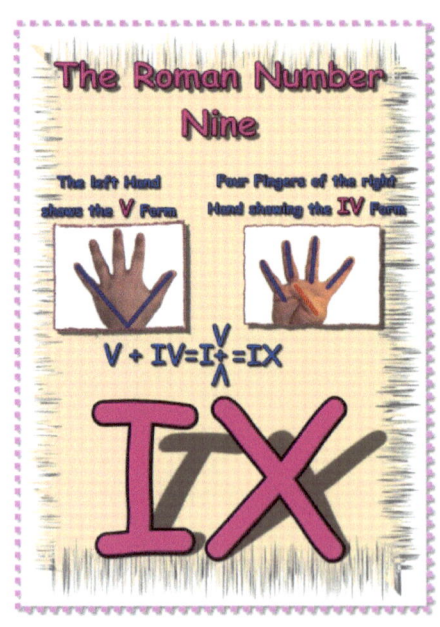

The Roman number Ten

For number **Ten**, one needed to show only both hands with the **ten** fingers. To make it easier, the Romans used the **X** as a Form for the quantity ten. To assume that, the **X** Form arose from the possibility of putting the symbol **V** for number **Five** together, so that the result is a form that represents **X**. Now the Romans recognized that the use of letters and symbols was an easy way to explain larger arithmetic and logic units, quantities and numbers. They decided to develop this symbol representation further and further. By showing the number five on both hands and then letting both hands meet in the intersection (see picture), the symbol **X** will be visible. This symbol or the letter **X** for the Roman number **ten** is still in use today.

Ten Pictures = X Units

The Roman number Eleven

With number **eleven** it is a little bit difficult, because it depends here on the concentration of the opposite person. One shows first both hands and directly afterwards One finger of One hand, therefore One shows first number Ten and additionally picks up One finger of One hand to indicate number One, and so would number **Eleven** be displayed.

Eleven Pictures = XI Units

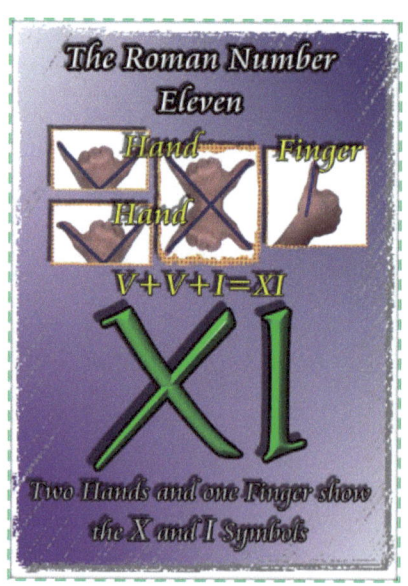

The Roman number Twelve

With number Twelve it is quite difficult but much the same as number eleven. One shows first both Two hands and afterwards the thumbs and index fingers of One hand; it means, first number Ten and then One picks up Two fingers additionally of One hand and so number Twelve is visible.

Twelve Pictures = XII Units

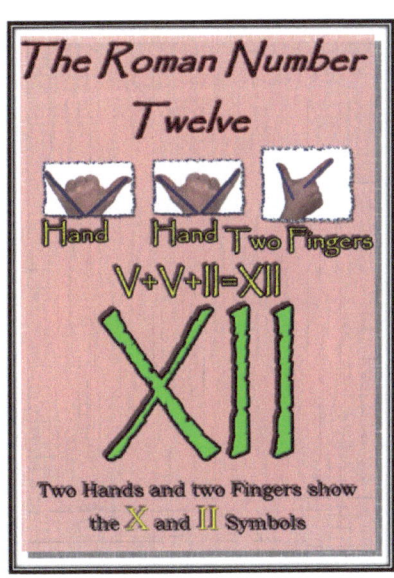

The Roman number Fifty

As the quantities were increased more and more, for the ancient users it was not possible to count only with the hands, this kind of representation (to show the quantity with the hands and fingers) was not sufficient any more.

Assuming that one carries heavy loads or several articles on his arm, how could he be able to use then the hands without losing the carried things?

Thus, the old Romans imagined that perhaps if one used both hands for number Ten, then he could use the whole free arm for the indication of number **Fifty**.

This was the first step to find one more symbol for the visible representation of larger quantities. The bended arm shows the form like the letter (**L**), and thus the symbol (**L**) signified the quantity of number **Fifty**.

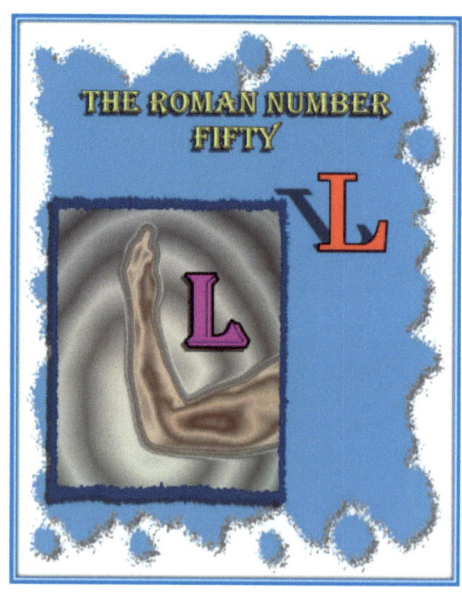

The Roman number One Hundred

If someone would like to carry or lift a large object, then he must lift those Units with both arms, i.e. he uses both arms at the same time. Both (**L's**) together build and show the form (**C**) as shown in the picture. We already know that the symbol (**L**) stands for number Fifty, so logically which means; that Two (**L's**) give in One Form (**C**) and now the number is called **One Hundred**.

Thus, it is fixed that the letter (**C**) receives the value of **One Hundred**. Each arm signifies the value **Fifty** and both arms show the **double quantity** thus the value is **One Hundred**. The Two arms form a symbol, which looks like the letter (**C**). Therefore, we also say nowadays for example (with full hands).

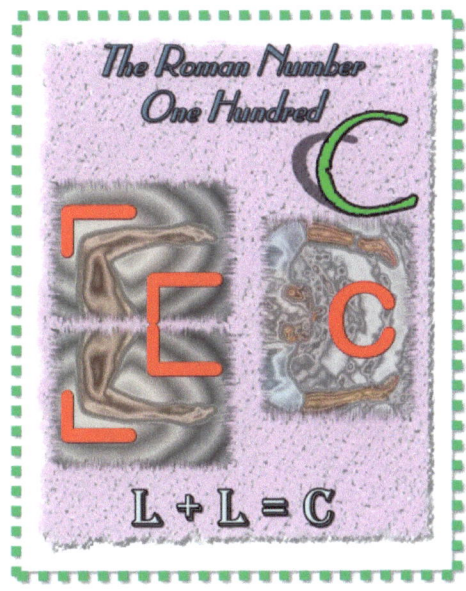

The Roman number Five Hundred

If someone would like to lift heavy weights or Units up from the floor, he does not only have to embrace this article, but he would also have to press it against his body and hold his hands together to raise it.

A new Form is born from that shape; it is the Form (**D**).

Both (**L's**) or closed (**C**) result in form (**D**).

This symbol obtained then the value (**Five Hundred**).

Hereby the Romans could then show a quite large quantity with only One letter form or symbol and use it for written mathematical equations.

This was a culturally big and simple step in order to replace large quantities with optical symbols.

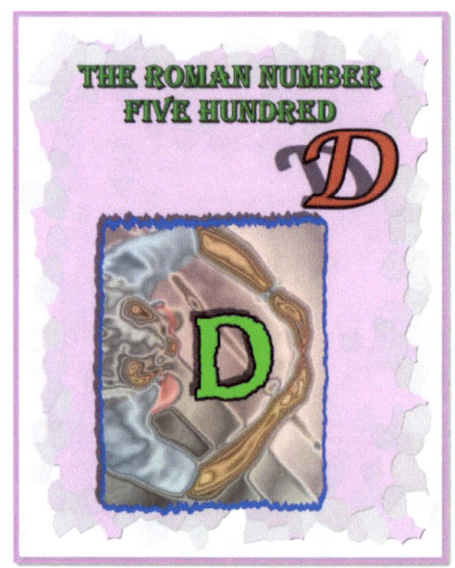

The Roman number One Thousand

As far as we know or better said according to the knowledge that we have today, the number (**One Thousand**) exists and also is represented as a Roman number in a symbol form. How could the number One Thousand be developed and which symbol could be used to display it?

What should I do if I would be no longer capable of lifting up a heavy load neither with hands nor with closed arms?

I think the only possibility would be to use my whole body for that. This means I can lift this heavy weight only through levering it up with my body. This is visible when watching a sport art like weightlifting; they could never move such enormous weights without incorporating the whole body. Therefore, as the loads became too heavy, then one had to use the whole body as a lifting machine, and thus the logical and optical number system of the Romans is ended.

With the whole body! What a beautiful idea!

What kind of form or which kind of symbol could be now used?

Actually, quite simple, it is the form or letter (**M**).

When we imagine that a Roman man before lifting a load. How does he stand there?

Most possibly by standing up straight and with both arms hanging parallel to the body, now we connect the lines (as seen in the diagram)" arm, body and arm again", this gives a clear picture which is easy for everyone to recognize resulting with a form that represents the form and letter (**M**).

Thus, the number One Thousand is probably the maximum optical and mathematical computable and calculated number, in which the Romans were able to do mathematical equations with?

It is like the power measurement in horse power.

I could say; the human power would be measured with the

67

M = ONE THOUSAND

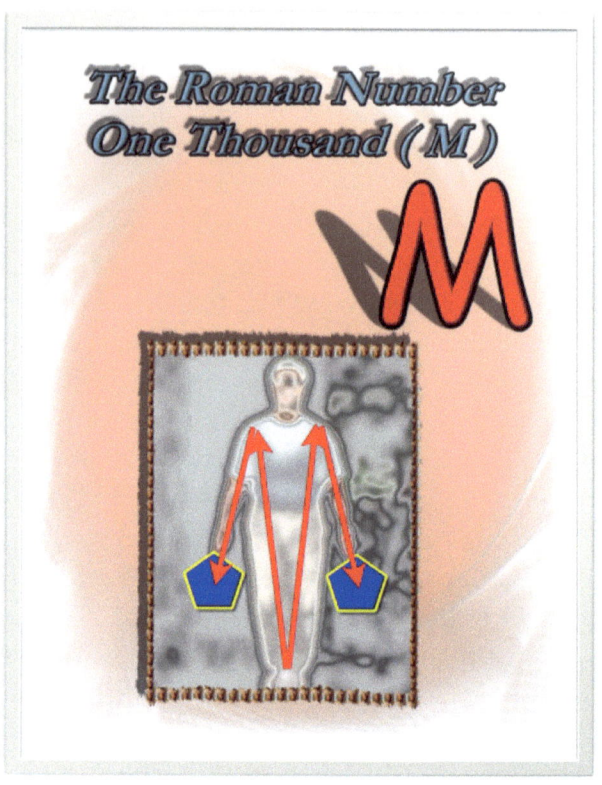

Chapter Two (2)
The Arabic Numbers

The ancient Egyptians and their Pyramids

By applying a simpler number system, it was surely possible and easier for the ancient Egyptians to create, design and build their pyramids with perfect precision due to simple mathematical calculations.

The Angle Principle and the Logic of the Arabic Numbers

Many thousands of years ago, there were people in a country called Egypt, they lived and still live around the river Nile, they were interested and fascinated with sculptures as descriptive representations. The ancient Egyptians were at that time very busy with mathematics, physics, medicine and astronomy as well as other scientific faculties.

Someday and somewhere in this old Pharaohs realm, that once was so large that it stretched across Asia and Africa, an "intelligent brain" had the idea to simplify all mathematical fields and the most difficult calculations with only One symbol or One Form.

This symbol consists of two lines, which were connected together at an intersection point resulting in an angle at the point of contact, that's all!

The Angle

The angle probably represents the idea, origin and importance of numbers and applied mathematics.
Nowadays the appearance of numbers looks similar, almost nearly, however not identical to the old Ones.
Their appearance and forms became only more beautiful, fine, round and smoother, but these modern numbers have nothing in common with the original styles and variations of the old numbers.
The old numbers became through their development their meanings, their sense and style and their reasons, why they were shaped this way.

Training example

As an example, I shall try to draw the form of the Arabic number (Two) using the principle of the angle.

Do some Exercises

Try to draw and to find out the original forms of the Arabic numbers using the principle of the angle before you continue to read on the next pages and compare it later on.

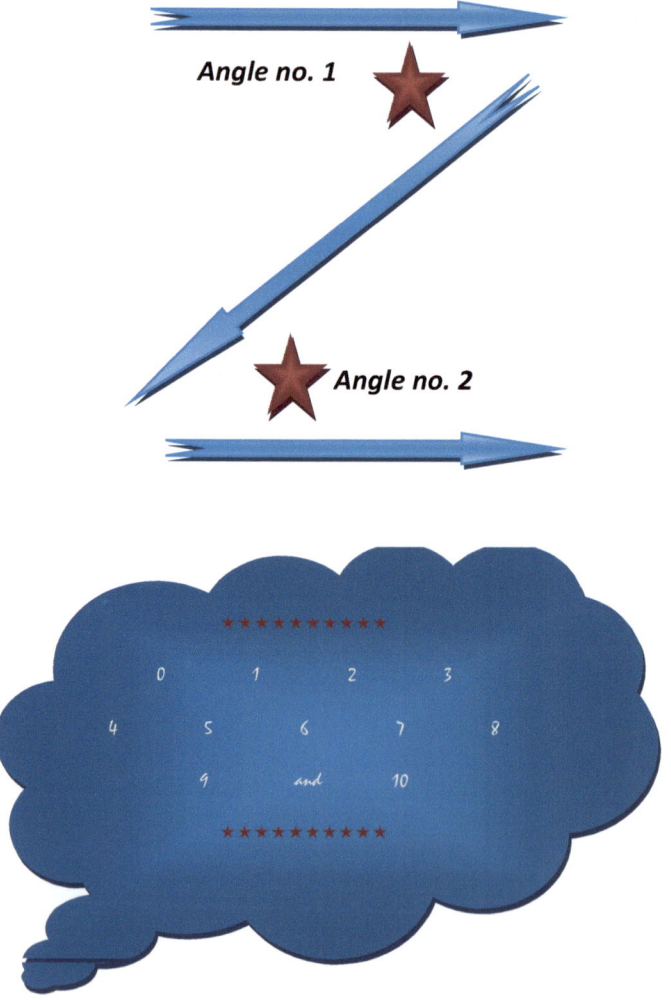

The Arabic number One

Until the Arabic number One (1) was sketched and used, the Roman number One (I) was probably applied because of its simple and uncomplicated form.

The intelligent Egyptian thought, for example, if he would like to find out and write down the quantity of the fish that he captured, what would be the simplest way to represent these without facing mathematical difficulties or complicated representations? The simplest solution is a form with an angle, as shown on the next illustration. Here we see an indication with only one angle. Therefore, it was to be used as a designation of one unit. From now on, the Egyptian starts counting with angles in one Form.

One Picture = 1 Unit

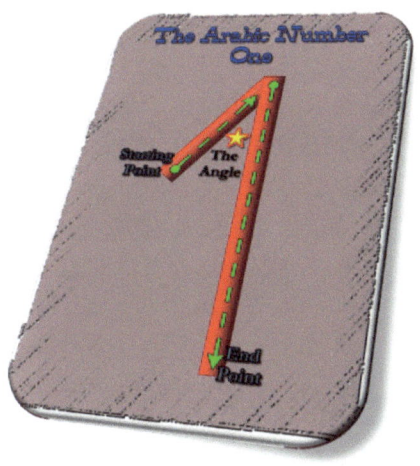

The Arabic number Two

Our Egyptian started thinking again and wondered and wanted to find out if one could continue with a small quantity, say two sheep. Which possibilities would he have for a simple representation of the two sheep, except to show it graphically?

There were Two solutions:

The first one: One uses the Roman Form (I) twice (II). With this method it would be possible to give even more forms and indications for much larger quantities.

The second one: relatively simple, he sketched a symbol with two angles, which is illustrated on this page.

Two Pictures = 2 Units

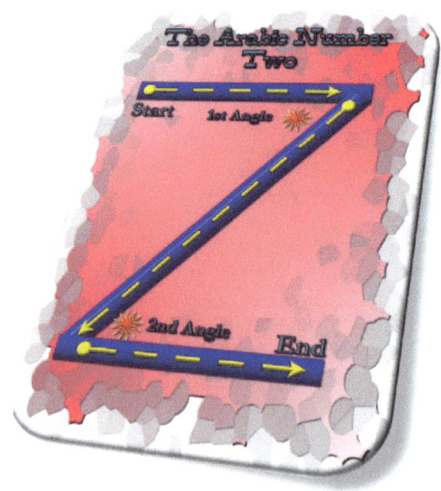

The Arabic number Three

It is pretty much the same like before, so he used the same method to determine three units.

It comes easier to engrave or write an indication with Three angles than engraving Three times the Roman number (I) e.g. (III), and thus this method of using this form for numbers came into the world.

Three Pictures = 3 Units

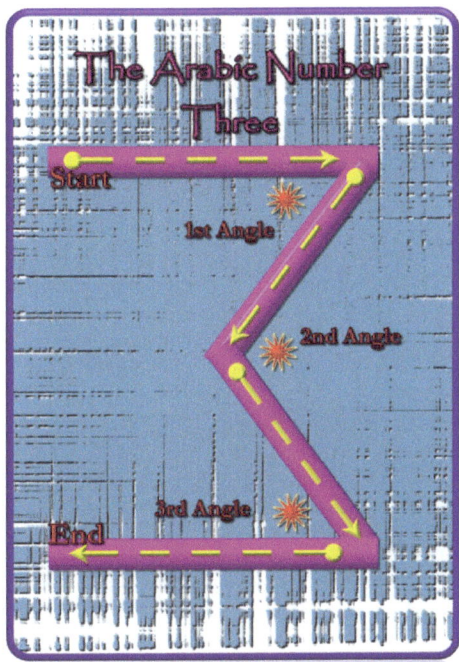

The Arabic number Four

Here, our intelligent Egyptian made up his mind to realize how to develop a smart and simple symbol, which could display Four units in One graphic form.

"These thoughts from our intelligent Egyptian turned out to be quite successful", or not!

Four Pictures = 4 Units

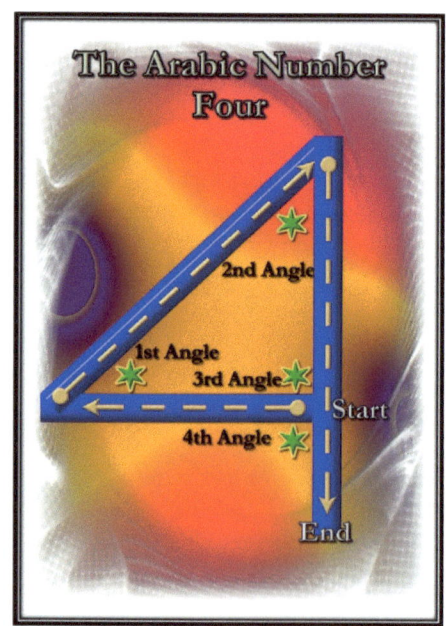

The Arabic number Five

Here, it became difficult and complicated to recognize and understand that the Romans used this Form (V) to display *Five Units*, or number five. Up to now our Egyptian friend was very experienced in broadening and developing graphic symbols for numbers, and so he came to develop this beautiful Form of quantity of five unites, later known to be number Five.

Five Pictures = 5 Units

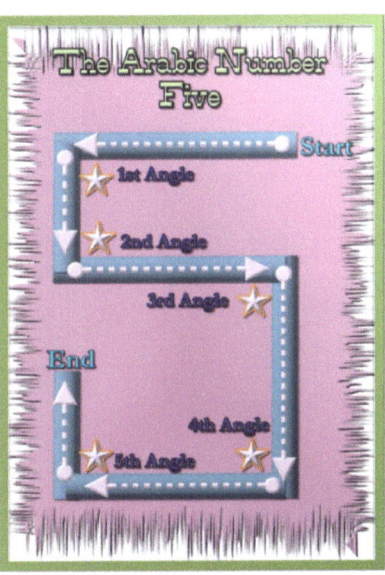

The Arabic number Six

The Roman numbers did not inspire our mathematician and architect anymore, so he continued to develop his own graphical number Forms and symbols in the same way. Here we look on at the symbol for number Six, which is simple but astonishingly intelligent.

Six Pictures = 6 Units

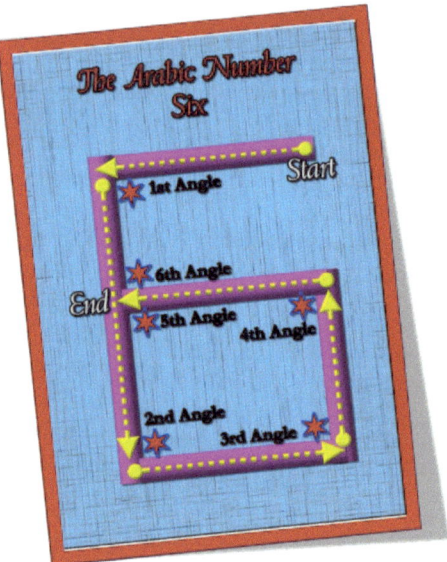

The Arabic number Seven

Our mathematician became very bored by looking at the Roman numbers, since he did not recognize any improvement in the Roman number Design. The number forms and symbols repeated themselves and showed no further imaginative power. So he therefore designed and developed a beautiful form to represent the quantity of Seven Units.

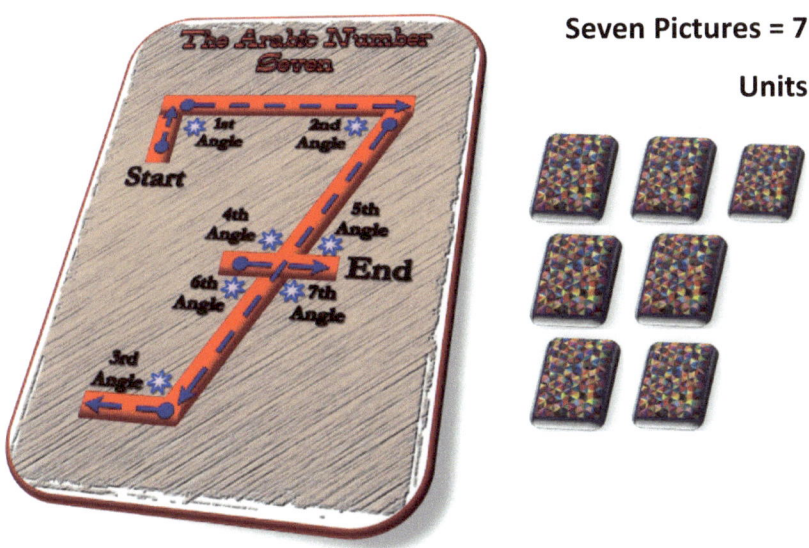

Seven Pictures = 7 Units

The Arabic number Eight " 1st Form

The good and curious philosopher had developed number eight systematic simply by closing the opening form of number Six with a dash so that it produced Eight angles, this was the first form of number Eight.

Eight Pictures = 8 Units

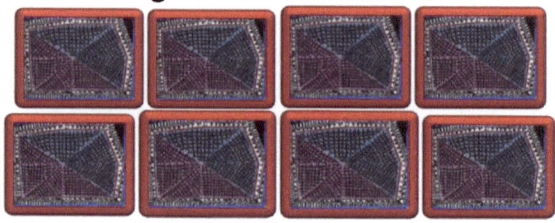

The Arabic number Eight " 2nd Form "

But our mathematician thought and wanted giving number Eight independence in its form; therefore, he has designed a new and beautiful form of number Eight (the second form).

Eight Pictures = 8 Units

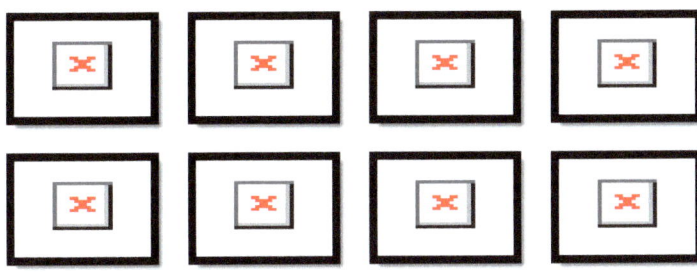

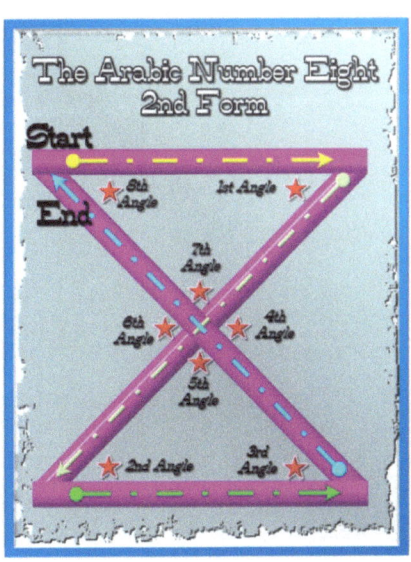

The Arabic number Nine

To develop further number forms, our good and intelligent Egyptian discovered another completely new way to represent the quantity of nine units. This new form was designed in such a way as is shown on the next illustration.

Nine Pictures = 9 Units

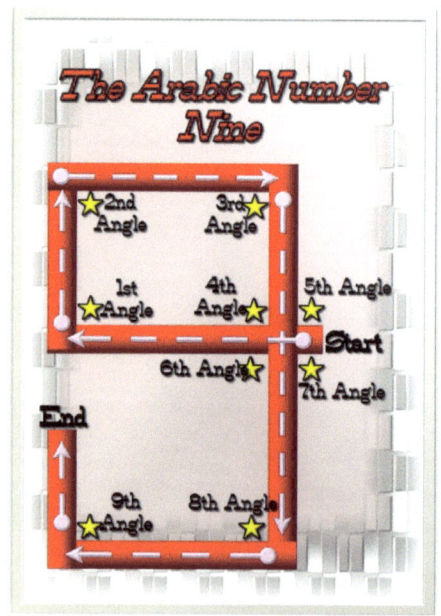

The Arabic number Zero

The number Zero was developed much later around 800 AD. Our old mathematician was in the meantime an Arabian, his name is **Alchwarazmi**. He wondered at the end of his creative era *"If I only have ONE unit, then I have a form with only One angle; but if I have __NOTHING__ at all, then this should also be displayable in a graphical form, which could clearly show that I have __NOTHING__".*
And so, with these thoughts, he concluded and developed the form on the following illustration.
N. B. The number Zero doesn't exist in the Roman number system (   ) NOTHING.

Where are the pictures?
Unfortunately, there are no pictures for **Zero**, because **(0)** means total Nothing or there's nothing to see!

Chapter Three (3) "Presentation of my theory"

- I find that the Roman number Ten (**X**), as a single form, is more precise than the Arabic number Ten (**10**), which consists of two already existing pre-defined and combined numbers (**One**) and (**Zero**).
- *Ones Series* (Individual units) in the Roman number-system consists of ten numbers (**I** to **X**); The <u>Roman</u> number (**X**) shows the form of a quantity of **ten** units, this number belongs to a group of (**I** to **X**).
- This **Ones Series** in the Arabic number-system includes only Nine Signs or <u>numbers</u> (**<u>1</u>** to **<u>9</u>**).
- In the Arabic number-system, we have taken the number Ten (**10**) as a transition and starting point in group is called group of Tens or **Tens Series**.
- The number Zero is not counted in the Ones group; so, we have only **nine** numbers which are considered in the regular Arabic counting-system.
- I took the numbers from One to Ten (with the **<u>new number Form of Ten</u>** and use this in an Ones Series) so that I could extend the group by Ten single number Forms instead of Nine numbers.
- The Arabic numbers from (**<u>0</u>** to **<u>9</u>**) were exactly defined, and every number was considered with a single form or symbol to easily recognize and determine the value of each number. The number Ten was forgotten to be defined.

- These symbols and forms of the numbers are of great importance for mathematical calculations. If I do not recognize the value of a number, how could I then build calculations or solve provided mathematical tasks?
- It seems to be the fact, however, what one has not taken the Arabic number counting-system into account, that the counting with both hands not with number Zero starts but with number One and ends with number Ten i.e. <u>One</u> ~ <u>Ten</u> but not with number Nine i.e. <u>1</u> ~ <u>9</u>.
- This is the <u>reason</u> why I´m trying to modify a new counting system by changing only one Form and inventing a new number.
- The second number group, which is called (**Group of Tens**), starts with the Arabic number **eleven** up to **One Hundred**.
- After I had finished writing the book in German and followed up the number forms and redesigned the forms of number (Ten), it was clear to me through this procedure that the Arabic number system is most probably incomplete or better said "not quite exact".
- The number ten should be only consisting of one graphical form with **ten** angles and not from predefined two number forms like (<u>One and Zero</u>).
- A possible approach that would be: systematically after number Nine One closes the number form (Nine) with a dash, or like the third example by adding a horizontal dash in the middle of Form Eight, and now very simply we obtain the new graphical symbol for number

85

(**Ten**) which gives a final result of the principle of angle counting from Zero to Ten.

My Conception, Imaginations and Calculations

My Symptom, (I know it could be confused):
1 x TEN = (ONE times TEN) = TEN, and
2 x TEN = (TWO times TEN) = twenty, and so on. However: 10 = 1 x 0 = (ONE time ZERO) = ZERO and not TEN!!! It is because ZERO stands after number ONE on the right-hand side, but if it stands at the left-hand side such as 01, it results in the quantity of ONE, these numbers together are equal to ONE (01=1), and ZERO has a logical meaning of a quantity of NOTHING. Therefore, I term this form (10) as;

The illogical form of the Arabic number ten

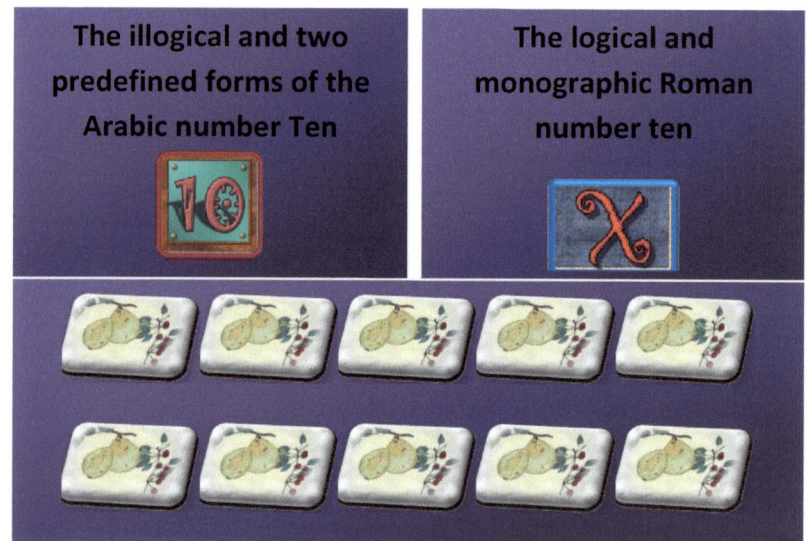

The Theory and the Logic of number Ten / The Idea

Approximately 3 years ago (2002), I inserted unconsciously and without explanation these symbols (!?) after the number **Nine**, such as (0123456789**!?**). As if I asked myself unknowingly: "Why does it not continue from here on? ". I mean after number nine, is there anything missing? Probably, I could not accept that the Logic of the Arabic numbers ends after the logical number **Nine**, and the number **Ten** does not exists as a unique form!

I was not very satisfied with the conclusion of my book without writing logical information and solving the problem concerning the number Ten.

All this left me with a big open question mark (?).

In November 2005, after I had logically worked on the whole book, I had an "**illuminating thought**" that the number Ten could look like the number Nine or Eight but with a small addition in the design that would represent the number Ten in One unique form and as a quantity term like that of the Roman number Ten (**X**).

For this form of the Arabic number Ten (10) "according to the angle principle" I did not find any logical explanation, because it consists of two predefined Forms the numbers One and Zero.

As a consequence, and according to the observance of the angle principle and the description of both the Roman numbers (I, V, X, L, C, D and M) and the Arabic numbers (0123456789), I have developed three new Forms for the Arabic number Ten and there I would like to suggest and present my new theory.

87

My new Arabic number Ten (1st form)

The first description of a new number form, which represents the logical meaning of the quantity of number Ten as a new Form; here we have an extension of the Arabic number Nine, which turns it into the new Arabic number Ten.

Here we see a form with ten angles, these are ordered so that one can represent ten angles in only one single form.

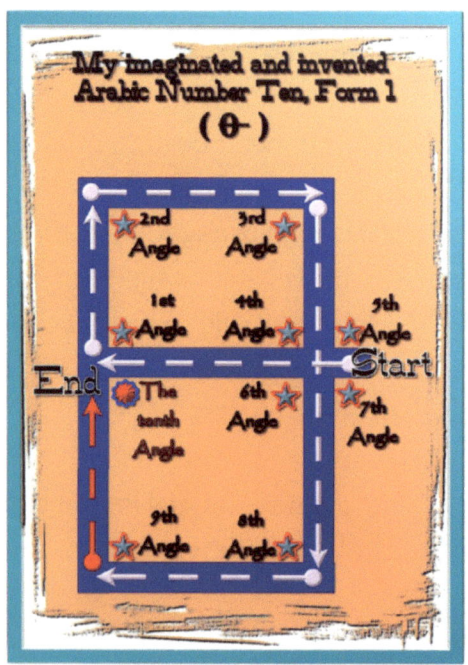

My new Arabic number Ten (2nd form)

It is the second possible extension of the Arabic number Nine which turns it into number ten. In addition, this form exhibits ten angles.

This form could be the most useable One for further mathematical equations. This design is a mirror image of the first form which is described on the previous page.

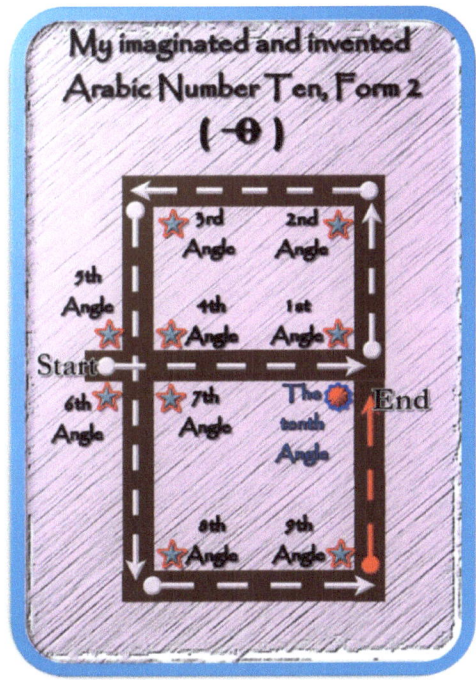

My new Arabic number Ten 3rd form (8)

This is the third and last possible concept to extend number eight into the new form of number ten.

We also see here the ten possible angles, which are clearly recognizable.

Now it must be decided, which of these Three forms or symbols offers the easiest usable and best handling for applications in mathematical processes, calculations and solving of arithmetic operations.

For me personally, the third and last form would be the best and easiest symbol to use. It is simple to understand and easy to represent and it is also different from the other forms or symbols. It is my favorite Form of the new Arabic Number Ten.

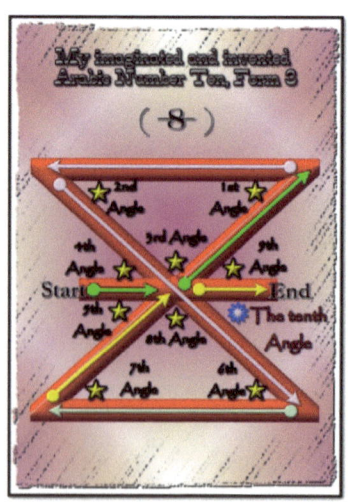

My new Arabic number Eleven (1 8 1)

Here I begin with my new numbering and computing system. In addition, my book ends here, so that I don't cause confusion and provocation to the young readers, because as I said these thoughts are only my own, concerning a change of the Arabic number system and they are not based on any registered scientific or mathematical research.

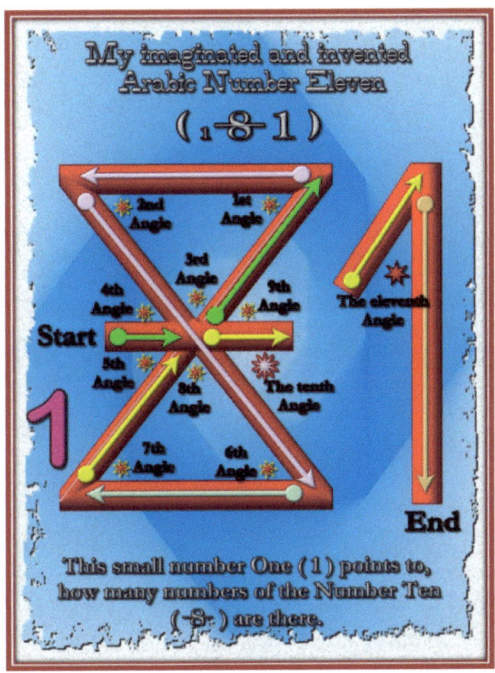

Modern day Counting system					
Zero Margin	One's Series	Ten's Series	Hundred's Series	Thousand's Series	Million's Series
0	1, 2, 3, 4, 5, 6, 7, 8, 9	10 to 99	100 to 999	10^3 to 999.999	10^6 and so on
Content of the Series	Here are only 9 Numbers	Here are 90 Numbers	Here are 900 Numbers	Here are 999000 Numbers	Etc.

My counting system					
Zero Margin	One's Series	Ten's Series	Hundred's Series	Thousand's Series	Million's Series
Zero	One to Ten	Eleven to One Hundred	Hundred One to One Thousand	Thousand One to One Million	Million One to … and so on
0	1,…,9 and ~1~0	~10~1 to ~10~2 or (~100~)	~100~1 to ~10~3 - or ~1000~	~10~31 to ~10~-6	~10~-61 to…
Content of the Series	Here are Ten Numbers	Here are 90 Numbers	Here are 900 Numbers	Here are 999000 Numbers	Etc.

Some Number Examples with my counting system

Today's counting system		My Counting system	
		By only using the 3rd Form of TEN (~~8~~)	By using the 2nd Form (~~0~~) together with the 3rd Form (~~8~~) of Ten
10^1	10	~~8~~	or (1~~0~~)
11	11	~~8~~-1	or (1~~0~~1)
99	99	~~8~~-9	or (9~~0~~9)
10^2	100	~~8~~-2	(~~8~~ ~~8~~) or (1~~0~~~~0~~)
101	101	~~8~~-2 1	(~~8~~ ~~8~~-1) or (~~8~~ ~~0~~1)
110	110	~~8~~-2 ~~8~~	(~~8~~ ~~8~~ ~~8~~) or (~~8~~ ~~8~~ 1~~0~~)
10^3	1000	~~8~~-3	or (~~8~~~~8~~~~8~~)
1001	1001	~~8~~-3 1	or (~~8~~~~8~~~~8~~ 1)
1010	1010	~~8~~-3 ~~8~~	(~~8~~~~8~~~~8~~ ~~8~~) or (~~8~~-3 1~~0~~)
1073	1073	~~8~~-3 7~~8~~3	(~~8~~~~8~~~~8~~ 7~~8~~3) or (~~8~~-3 7~~0~~3)
10^6	1000000	~~8~~-6	or ~~8~~~~8~~~~8~~~~8~~~~8~~~~8~~
1000001	1000001	~~8~~-6 1	or ~~8~~~~8~~~~8~~~~8~~~~8~~~~8~~ 1
And so on			

Chapter Four (4)
Muslim Scientists and Mathematicians of the Middle Ages

1. **Alchwarazmi**, the "father of Algebra":
   This explanation about this mathematician and the development of the number **Zero** is based on research, activities, which I accomplished for the production and technical contents of my book. The invention of number **Zero** and the systematization of counting, we thank an Arab, who lived (around 800 AD) and helped in advancing and simplifying mathematics, his name is:
   Abu Dschafar Muhammad ibn Musa Alchwarazmi
   One assumes that he wrote Two mathematical works; One of them a book on **arithmetic**, the other one on **algebra**. In English, it begins with the linguistic usage of the word "Algorithmic digit", expressing a mathematical term, which nowadays plays a major role in mathematics and computer science, which we know today as *The Algorithm*, he was indeed the first who specified the systematic computing (arithmetic) procedures for counting numbers and also the problem of solving equations that we know as algorithms. Starting from this time, there is a clear definition between the number writing of the Romans and of the Arabs.
   Unfortunately, his exact date of birth as well as death is not well-known (around 770 and 850 AD). His family left Iran shortly after his birth and went to Bagdad, Iraq.
   He also helped to bring "Arabic numbers" into use into the Islamic world, as well as later (around 1200 AD) into Europe. He also demonstrated operations with fractions for the first time. He influenced the growth of science

and mathematics. Several of his books were translated into many other languages, and were used as university textbooks until the 16th century. His approach was systematic and logical. He brought together the knowledge of his time on various branches of science, especially mathematics, and also added his original contributions.

2. **Omar Al-Khayyam** (about 1045 - 1125 AD): Another great Muslim mathematician was Omar Al-Khayyam. He is best known today for his poetry, but his contribution to mathematics was also great. He showed how to express roots of cubic equations by line segments obtained by intersecting conic sections. Al-Khayyam was an outstanding poet, mathematician, and astronomer. His work on algebra was known throughout Europe in the middle Ages, and he also contributed to a calendar reform. He refers in his algebra book to Pascal's triangle.

3. **Al-Khashy** was born in 1390 AD in Kashan, Iran and died in 1450 in Samarkand (now Uzbekistan). He calculated 1 (PI = Π) to 16 decimal places which was the best until about 1700 AD. He considered himself "the inventor of decimal fractions". He wrote *The Reckoners' Key* which summarizes arithmetic and contains work on algebra and geometry.

4. **Al-Biruny** (973 - 1048 A.D.) was a philosopher, astronomer, botanist pharmacologist, geologist and mathematician. He translated Euclid's work into Sanskrit language, and calculated the earth's circumference and radius with an accuracy that is close to today's measurements.

5. **Nasser Al-Din Al-Tusi** (about 1200 - 1275 AD) He can be called the *"Father of Trigonometry "*, pioneered spherical

trigonometry which includes six fundamental formulas for the solution of spherical right-angled triangles. One of his most important mathematical contributions was the treatment of trigonometry as a new mathematical discipline. He wrote on binomial coefficients which Pascal later introduced. He was also an astronomer, philosopher, and medical scholar as well as a mathematician.

Demolition and Destruction of Books, Libraries and civilizations

- The first fall of Baghdad (about 1260 AD), led to the end of the Abbacies Monarchy. About more than One million Muslims were killed and wiped out in Iraq. The major scientific schools, institutions, laboratories and even roads and waterways in leading Muslim centers of civilization were destroyed. The second fall was at 2003 AD through USA and its alliances.
- Another wave of hate and destruction came with crusades against Muslims in the Middle East, led by European Christian Kings, and also when the Christians took over Spain in 1492 AD. More than One Million volumes of Muslim works (Books and documents) on science, arts, astronomy, philosophy and culture were burnt in the streets in Granada.
- The Islamic culture was unique in its magnificence and learning.
  It was indeed a golden Are of about 800 years! and
- The Egyptian civilization lasts of a period of more than 3000 years until beginning of Greeks (Alexander; the Great!!), Romans (Cesar) and Christian's crusades.
- Some Authors or scientists wrote about the Arabs, showing them like Dummies or Bridges; that they were only there just to transfer and translate most of scientific fields from India or China to Europe! Or such as; if the

- Arabian and the Egyptian cultures and Civilizations wouldn't be existing.
- I think it is a kind of hate to them and to the Islam; this hate appears in their war-attacks or Crusades in the last centuries and also nowadays.

Thanks

Nevertheless, we can say that our Arabic numbers had a quite old and interesting journey, in their development and discovery. Finally, however they became widely accepted in relation to the complicated symbol system of the old Romans because of their simplicity and in the execution and way of writing them. Therefore, not only the simplicity in the way of writing is crucial for its use but also the simplicity of its symbols was a major reason why they were so wide spread.

In the Arabic numbers there is no confusion in recognizing different symbols, but with the Roman number system confusion was unavoidable for some people, for example the number symbols **IV** and **VI**. The Arabic system was easy for everyone to define e.g. the numbers **4** and **6**.

This development of the number system was trailblazing for our advancement.

Without the possibility of an optical representation for the number **Zero**, it would not be possible to use or invent computers until today, and without computers our life would never had reached the standard of development that we now have today.

The numbers today became more beautiful in their execution and their Design, and although they look more beautiful in their design they have never lost their important meaning for today's science & Mathematics.

For the normal human being it is unusual to think about their origin or why and when the numbers were developed, and to that point, who modified and designed them at all?
We just take for granted that the numbers are there, because we use them in everyday situations and we could not imagine a life without them.
It only remains to say to these smart old Egyptians and Arabs, thanks for your efforts in representing mathematical procedures that made it as simple as possible for us. Not least they gave us the possibility to be able to develop and lead our cultural life which we live and lead today.

We have won now the advantage to look closely and to learn more about our Arabic number system, and who it probably developed, and how it was kept till our present time, and why we use them in modern day life. And because of these number symbols, it is very easy to represent sizes and measurements visually in order to do mathematical calculations.

Without the intelligent brains of the old Nations, the development of our lives as we know them today would certainly look very different than what we have today.

The development of mathematics made it also possible for us to discover a variety of other scientific fields such as physics or chemistry, which would not exist without formula evaluations. Therefore, one can actually see that because those intelligent people brought mathematics into our lives, a large step was done towards cultural developments and science, and because of this big step for mankind given to us from the old Egyptians is it now possible for us to study other dimensions, for example, the study of the universe.

Professor Kotomoto was very impressed, oh my god, he said.

I have to go to sleep, it was a very imagination night.

End of The Part one

www.ingramcontent.com/pod-product-compliance
Lightning Source LLC
Chambersburg PA
CBHW040317220526
45473CB00009B/2470